How Did You Count?

PICTURE BOOK

CHRISTOPHER DANIELSON

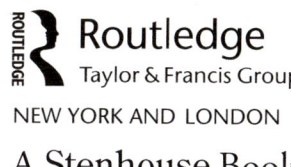
Routledge
Taylor & Francis Group

NEW YORK AND LONDON

A Stenhouse Book

Cover and interior photography by Asha Belk

First published 2025
by Routledge
605 Third Avenue, New York, NY 10158

and by Routledge
4 Park Square, Milton Park, Abingdon, Oxon, OX14 4RN

Routledge is an imprint of the Taylor & Francis Group, an informa business

ISBN: 9781625312938 (set)
ISBN: 9781032898353 (hbk)
ISBN: 9781003581482 (ebk)

DOI: 10.4324/9781003581482

Typeset in Gotham
by KnowledgeWorks Global Ltd.

Access the Support Material: https://www.routledge.com/9781032898353

For Product Safety Concerns and Information please contact our EU representative *GPSR@taylorandfrancis.com* Taylor & Francis Verlag GmbH, Kaufingerstraße 24, 80331 München, German
Printed in US at IBI

This is a book about counting, but not about right and wrong answers.

There are lots of interesting things to count. More important, there are lots of interesting ways to count them.

Once you know how many there are, count them in another way.

Turn the page to see what that means...

How many tangerines?
How did you count them?

Did you count the tangerines as four columns of three tangerines each?

Maybe you saw three zigzags of four tangerines.

Or two groups of six, or maybe you counted them one-by-one.

What other ways can you count the tangerines?

How many bowling pins?
How did you count them?

How many pudding cups?
How did you count them?

How many eggs?
How did you count them?

How many pots
of purple paint?

How did you count them?

How else could you count
them?

Can you find groups of four,
and five more?

How many basketballs?
How did you count them?
Did you count any basketballs
that you cannot see?

How many soccer balls?
How did you count them?

How many apples?
How did you count them?

Now how many apples?
How did you count them?

How many shoes for
how many children?
How did you count them?

How many flowers?
How did you count them?

How many grapes?
How did you count them?

How many dots are on
the tops of all the dice?
How did you count them?

Now how many dots are on the tops of all of the dice? How did you count them?

What will you count next?

How will you count them?

I made this book to spark conversation, thinking, and wonder.

I hope this is a book you will leave open, think about, and return to often. I hope you will share it with others.

I hope you will find interesting things to count in your world, and interesting ways to count them.

I hope you will send me a picture or drawing that invites many answers to the question "How Did You Count?"

Find me at talkingmathwithyourkids.com